T0092868

Change Request Impacts in Software Maintenance

Computational Intelligence and Management Science Paradigm

Series Editors:

Vijender Kumar Solanki and Raghvendra Kumar

This book series will focus on the latest concepts of Machine Learning, Data Science and many more emerging technologies. It intends to focus on short form books on cutting edge research methodology, concepts, and applications in the field of computational intelligence and management sciences and include interdisciplinary emerging trends. The main focus for this new series is on the development and application of emerging technologies and applications for extracting knowledge from structured and unstructured data. Authors and editors from academic and industries worldwide will be invited to submit book proposals relevant to emerging computational intelligence and management sciences concepts technology, applications and case studies.

Change Request Impacts in Software Maintenance

Madapuri Rudara Kumar, Annaluri Sreenivasa Rao, Kalli Srinivasa Nageswara Prasad and Vinit Kumar Gunjan

For more information about this series, please visit: https://www.crcpress.com/Computational-Intelligence-and-Management-Science-Paradigm-Series/book-series/CRCCIMSP

Change Request Impacts in Software Maintenance

Madapuri Rudra Kumar,
Kalli Srinivasa Nageswara Prasad,
Annaluri Sreenivasa Rao and
Vinit Kumar Gunjan

CRC Press
Taylor & Francis Group
Boca Raton London New York

CRC Press is an imprint of the
Taylor & Francis Group, an **informa** business

First edition published 2020
by CRC Press
6000 Broken Sound Parkway NW, Suite 300, Boca Raton, FL 33487-2742

and by CRC Press
2 Park Square, Milton Park, Abingdon, Oxon, OX14 4RN

© 2021 Taylor & Francis Group, LLC

CRC Press is an imprint of Taylor & Francis Group, LLC

Library of Congress Cataloging-in-Publication Data

Names: Kumar, Madapuri Rudra, author. | Prasad, Kalli Srinivasa Nageswara, author. | Rao, Annaluri Sreenivasa, author. | Gunjan, Vinit Kumar, author. Title: Change request impacts in software maintenance / M. Rudra Kumar, Kalli Srinivasa Nageswara Prasad, Annaluri Sreenivasa Rao, and Vinit Kumar Gunjan. Description: First edition. | Boca Raton, FL : CRC Press, 2020. | Series: Computational intelligence and management science paradigm | Includes bibliographical references and index. Identifiers: LCCN 2020013900 (print) | LCCN 2020013901 (ebook) | ISBN 9780367898748 (hardback) | ISBN 9781003032564 (ebook) Subjects: LCSH: Software maintenance--Management. | Computer software--Management.Classification: LCC QA76.76.S64 K86 2020 (print) | LCC QA76.76.S64 (ebook) | DDC 005.1/6068--dc23 LC record available at https://lccn.loc.gov/2020013900 LC ebook record available at https://lccn.loc.gov/2020013901

ISBN: 978-0-367-89874-8 (hbk)
ISBN: 978-1-003-03256-4 (ebk)

Typeset in Times
by Deanta Global Publishing Services, Chennai, India

CONTENTS

PREFACE

Versioning is mandatory for software systems that enable improvi
aation, simplification and customization to fit different application
scenarios in the context of target users or organizations. The bug
fixing and software reusability are other contexts that deserve effective Impact Analysis of the changes required for the source software
sources. The unexpected complications appear due to changes applied
against the change requests, often ruining the product credibility and
brand reputation. Hence, the need for performing change request
Impact Analysis is obvious for software maintenance, versioning,
software reusability and bug fixing. The efforts behind the framing of
chapters in this book endeavour to give a bird's eye vision on different
factors of change request Impact Analysis in software versioning and
maintenance that include software reusability and bug fixing.

The overall content of the book delivers essential concepts to
diversified readers from the domains of software engineering
research and requirements management, the industry of
software maintenance and graduate and post-graduate students
of software engineering.

AUTHOR BIOGRAPHIES

Madapuri Rudra Kumar holds a PhD in Computer Science from Jawaharlal Nehru Technological University, Anantapur, Andhra Pradesh, India. Presently, he is working as a Professor and Head of the Department of Computer Science and Engineering, Annamacharya Institute of Technology and Sciences, Rajampet.

He has 16 years of teaching experience. He has published and presented papers in national and international journals, conferences and seminars. His research interest includes Software Engineering, Data Mining, Machine Learning, etc. He is a member of various professor bodies.

Kalli Srinivasa Nageswara Prasad is a Professor of the Department of Computer Science and Engineering in GVVR Institute of Technology, Bhimavaram, affiliated to Jawaharlal Nehru Technological University, Kakinada, Andhra Pradesh. He received his PhD in Data Mining from Sri Venkateswara University, Tirupathi. He obtained his MTech degree in Computer Science and Engineering from JNTUK, Kakinada; MS degree in Software Systems from BITS-Pilani; and AMIE in Computer Science and Engineering from Institution of Engineers, India. He has more than 20 years of teaching experience. He has published around 21 research papers in reputed national and international journals and has attended many workshops and conferences. He has guided many students in their projects and theses. His research interests are Data Mining and Big Data. He is a life member of ISTE and IEI.

Annaluri Sreenivasa Rao holds a PhD in Computer Science from Sri Venkateswara University. Presently, he is working as an Assistant Professor of the Department of Information Technology, VNR Vignana Jyothi Institute of Engineering and Technology, Hyderabad. He has 20 years of teaching experience. He has published and presented papers in national and international journals, conferences and seminars. His research interest includes Text Mining, Machine Learning, Software Engineering, etc. He is a member of various professor bodies.

Vinit Kumar Gunjan is an Associate Professor of the Department of Computer Science and Engineering at CMR Institute of Technology, Hyderabad (affiliated to Jawaharlal NehruTechnological University, Hyderabad). He is an active researcher; has published research papers in IEEE, Elsevier and Springer conferences; and has authored several books and edited volumes of Springer series, most of which are indexed in SCOPUS database. He has been awarded the prestigious Early Career Research Award in the year 2016 by Science Engineering Research Board, Department of Science and Technology Government of India. He is a Senior Member of IEEE and an active Volunteer of IEEE Hyderabad section, and has volunteered in the capacity of Treasurer, Secretary and Chairman of IEEE Young. Professionals Affinity Group and IEEE Computer Society. He was involved as an organizer in many technical and non-technical workshops, seminars and conferences of IEEE and Springer. During the tenure, he had the honour of working with the top leaders of IEEE and has been awarded best IEEE Young Professional award in 2017 by IEEE Hyderabad Section.

1

INTRODUCTION

The function of Impact Analysis is to estimate the outcome prior to the implementation or post-implementation of a cluster of modifications on the software [1,2]. In case the prediction is performed prior to system modification, this study aids software developers to evaluate the affected software components. This is also referred to as the estimated Change Impact Analysis. In contrast, in the case of predicting an output followed by significant changes, this analysis suggests program testing managers execute regression test analysis, by minimizing the count of executable test cases to the ones which are affected by the proposed modification.

To perform Change Impact Analysis, various methods and implementations [3] are required for diversified software stages like maintenance and development phases. In particular, Change Impact Analysis techniques [4–7] are commonly used in software maintenance level. Here, a class remains as a significant driver for analysis. On the other front, the drawback in the development stage is that "all classes are not fully developed", as the program is yet in the development stage [4]. Thus, existing Change Impact Analysis techniques are not applicable for development phases.

Earlier Change Impact Analysis methods, which are utilized for improvement phases, include incompletely developed classes among the artefacts. This is one of the significant threats to Impact Analysis. Hence, this research work evaluates an efficient analysis of the abilities of traditional Change Impact Analysis methods. This proposed work classifies efficient and poor capabilities of existing Change Impact Analysis techniques to aid in the optimal execution process of Change Impact Analysis for the improvement stage.

This book is structured in such a way that Chapter 2 explains the nomenclature and Chapter 3 explains similar work. In Chapter 4, we elaborate on the review criterion. Chapter 5 depicts the efficiency of existing Change Impact Analysis techniques and introduces a different Change Impact Analysis method for the improvement phase of the software. In Chapter 6, the research work is concluded.

The primary aim of Change Impact Analysis is to detect unexpected or ripple effects of suggested software modifications. It determines the possible effects of the modification or predicts the software modules or components to be altered to trace out a change [8]. In particular, Change Impact Analysis plays a crucial role in executing multiple functions such as estimating development cost, project overheads, handling modifications, system strength or detecting errors in a system.

The significance of Change Impact Analysis is to recognize work products which are threatened by modification, desired time and change cost of the components which can be impacted by change [9]. Generally, the impact of modification relies on the whole volume of change; i.e., the small project faces less effect, and the impact is higher on big projects [10].

However, multiple factors impact the reason behind the change proposal. For instance, some of the most observed causes include bug removal, fault correction, feature additions, exposing the project to an unknown environment and quality enhancement. These modifications often come with associated risks such as unintended side-effects, unforeseen errors and malfunctions. At times, choosing the best Impact Analysis remains a tough task as choosing such technique might restrict the program from making possible decisions at the right time.

In addition, few potential challenges and specific issues still exist during the implementation of Change Impact Analysis. Thus, handling modifications faces multiple threats extending from detecting the modification necessary to evaluate the possible impact of the suggested change across the project to executing the planned and related modifications. In addition, it also includes the efficient handling of the storage of previous versions of the program and obtaining all the information associated with the proposed change [8].

Hence, the research work depicts Change Impact Analysis techniques and detects the potential experiments of Change Impact Analysis techniques to aid software projects through guidelines.

As such, this book explains the utilization of Change Impact Analysis techniques in developing a software project and evaluates a new method of Change Impact Analysis procedure. Accordingly, on the basis of the research work, initial data gathering has been executed with the help of a project practitioner, and the resulting analysis is summarized qualitatively.

2

THE NOMENCLATURE

2.1 IMPACT ANALYSIS (CHANGE IMPACT ANALYSIS)

Change Impact Analysis or Impact Analysis is described as "a procedure of analyzing significant unexpected effects of a system or predicting what should be altered to realize a system change" [1]. In particular, Change Impact Analysis functions are mainly aimed at identifying various artefacts of software such as requirement component, design component, class component and test component. A system change adversely impacts these components/artefacts. Generally, a system change is caused by an effect of addition, deletion and alteration to existing or novel software artefacts. Based on data from artefacts that are likely to be impacted, an efficient strategy can be formulated on the sequences of steps involved in implementing the proposed modification.

Change Impact Analysis comprises two important approaches: dependency and traceability impact analyses. Dependency Change Impact Analysis is often referred to as program analysis. This type of Change Impact Analysis is primarily engaged in describing various associations between class artefacts by analyzing the internal source code's structure [1]. In particular, dependency Change Impact Analysis determines distinct source code elements which are likely to be affected by a system change. Researchers introduced multiple techniques of dependency Change Impact Analysis like control and data dependency impact analyses [11]. Control dependency Change Impact Analysis employs the conditional structure of a software program for executing the procedure and variable of that program which is analyzed through data dependency Change Impact Analysis.

Accordingly, the traceability Change Impact Analysis is used to analyse the associations between various artefacts across distinct phases of software. Due to this reason, traceability Change Impact Analysis approach is highly employed by researchers to execute Change Impact Analysis function in the development stage of software [12,13]. The only thing which differentiates it from dependency Change Impact Analysis is that traceability Change Impact Analysis techniques focus on interdependence of specified software artefacts in numerous software phases rather than on a single artefact.

Traceability Impact Analysis is categorized into two types: pre-traceability and post-traceability Change Impact Analyses [14]. The pre-traceability analysis caters an efficient mechanism to check whether all requirements are depicted in a recognized need specification format or not. On the other hand, the post-traceability Change Impact Analysis offers an effective mechanism for analyzing the execution procedure of all such requirements present in the specification document. It depicts the step-by-step execution process of requirements in the project.

Most of the research work on Change Impact Analysis is restricted to source code analysis [3–6] by utilizing the dependency Change Impact Analysis method. However, the analysis through source code is unable to detect all affected artefacts in the project.

Software artefacts including test and design need to be updated as per the system change. This ultimately depicts that these artefacts form an integral portion of those affected by the proposed modification. Hence, to analyze the frequent effects of a system change in a project, an efficient grouping of traceability and dependency impact analysis is required.

2.2 IMPACT ANALYSIS (CHANGE IMPACT ANALYSIS) PROCESS

According to recently proposed works, the impact is defined as an effect of one object on another. It is often described as the possibility of a system change [15]. The Change Impact Analysis determines the scope of modification demands as a source to achieve accurate resource development and scheduling and also verifies cost rationalization [15]. This process predicts a change in a specified project and also maintains relevant documentation if any change to that project is observed.

Change Impact Analysis techniques are mostly utilized for planning, implementing and accommodating specific project modifications, and observing the post-implementation impacts [16]. However, when a particular modification is observed in a project, it is significant to employ the suitable Change Impact Analysis method for analyzing the effect of a change. Selection of appropriate Change Impact

Analysis process allows a developer to detect the possible effects of an observed change in a system before the execution of those alterations. Thus, a suitable Change Impact Analysis method enables to perform an optimal detection process of changes.

Accordingly, in a few cases, Change Impact Analysis plays a significant role in planning software release and maintenance procedure so as to switch proposed modification into a system modification specification. In addition, in [17], Pfleeger proposed a study on computing numerous threats which are linked to a system change. In particular, this proposed work predicts specific consequences on resources, execution plan and efforts [17]. Accordingly, the underlying concept of all the definitions is that Change Impact Analysis enables further insights into the suggested change and does not alter any part of the project [18].

In [19], the three significant steps in a Change Impact Analysis procedure are explained:

- Evaluation of modification requirements and specific artefacts
- Determination of potential consequences
- Execution of necessary system modifications

The initial step of Change Impact Analysis is engaged in analyzing the affected software artefacts set assumed to be first impacted through the modification specification. The principal set of affected software artefacts is referred to as SIS set. In the second step, the Change Impact Analysis process determines the possible effects across SIS by eliminating the false software artefacts to extract unique artefacts. The resulting set is called CIS set or often called EIS set. The final step of Change Impact Analysis selects a group of affected software artefacts which is already altered while executing the actual system change. The resulting set of affected artefacts is represented as AIS set.

The procedural structure of a Change Impact Analysis is an iterative method. Through the execution of actual system change, newly affected software artefacts are identified along with the procedure. These newly explored artefacts are often referred to as DIS set. The DIS is a lower-estimate set of CIS.

As not each and every software artefacts in CIS is in the actual set, certain false software artefacts in CIS are referred to as the false positive impact set (FPIS). The FPIS is represented as an impact set which is over-predicted in CIS. The amalgamation of both CIS and

DIS artefacts and the removal of FPIS artefacts depict the results of AIS. That is, AIS = CIS + (DIS) − FPIS.

There are numerous parameters employed for evaluating CIS accuracy. This CIS accuracy is generated by the Change Impact Analysis procedures [20]. These parameters are primarily aimed at computing the extreme proximity between the results of AIS and CIS. If the results among CIS and AIS are closer, then a high rate of accuracy will be observed. For instance, in [20], the authors utilize three significant prediction parameters to estimate the efficiency of CIS results generated by Change Impact Analysis techniques. The three core metrics include comprehensiveness, accuracy and the value of Kappa (KV) [21].

The parameter correctness includes the ratio of estimated to total original impacted classes. The completeness parameter is used to determine the proportion of actual to total evaluated impacted classes. Finally, KV is employed for representing the agreement level between CIS and AIS.

2.3 CHANGE IMPACT ANALYSIS (CHANGE IMPACT ANALYSIS) TECHNIQUES

The section below depicts the different research studies conducted on Change Impact Analysis in contemporary literature. Most of the studies related to our research are presented below. The following sections in this chapter focus on describing proposed approaches, evaluating their regular procedures and also predicting them based on the capability of aiding multi-perspective Change Impact Analysis. This work focused on three primary requirements to execute an efficient Change Impact Analysis procedure:

- The ability to evaluate the diversified software artefacts
- Extending assistance to developers to identify the extent and areas affected by the change
- Assistance for multiple types of modification functions

2.4 DEPENDENCY ANALYSIS

This chapter depicts numerous Change Impact Analysis techniques using dependency associations between artefacts for executing actual Change Impact Analysis. These techniques are further classified into five different groups, which are explained in the sections below.

2.4.1 Distance-Based Graph Analysis

Initial studies developed in the Change Impact Analysis concept relied on dependencies of the program and used the related dependency map for predicting the change proliferation to different artefacts on the basis of proximity factors. However, these proliferations among dependencies resulted in an impact of the total system [22]. Accordingly, the effect calculated for a specific modification was often over-predicted and was not usable for re-engineering or maintenance of projects. Hence, the concept of dependencies was redesigned to restrict change proliferation to a pre-fixed length in the graph. This extended concept of dependencies presumes that all modifications affect only locally, and therefore unnecessary extra change proliferation is removed after the pre-fix distance is reached [22].

However, these techniques face challenges regarding the selection of appropriate transmission distance, which remains an unanswered task [23]. In case the distance is long, several false-positives can occur across the project as all the artefacts are regarded as being affected. On the contrary, a small distance can lead to several impacts being ignored as all potential artefacts could not be covered. The advantages and drawbacks of distance-based graph analysis are represented in Table 2.1.

2.4.2 Message Dependency Graphy

A unique type of dependency Change Impact Analysis method is proposed and designed for broadcast and event-driven systems. This approach analyzes the textual communication among the system's remote components. Later, through maintaining track of transmitted messages and documenting the observed transmission routes, the related dependency graph will be created. This graph is used for further impact analysis. Instances related to this work are explained in [24]. The advantages and drawbacks of message dependency graph analysis are represented in Table 2.2.

TABLE 2.1
Benefits and Drawbacks of the Distance-Based Graph Analysis Approach

Benefits	Drawbacks
Multi-perspective analysis possible	Difficult to define the correct distance
Understand and implement easily	No distinction among diverse modifications

TABLE 2.2
Benefits and Drawbacks of the Message Dependency
Graph Analysis Approach

Benefits	Drawbacks
Distributed systems analysis	Needs software execution
Event-based systems analysis	Needs message monitoring

TABLE 2.3
Benefits and Drawbacks of the Call Graph Analysis Approach

Benefits	Drawbacks
To extract the call graph easily	Only relevant on source code
	No distinction among diverse changes
	Granularity restricted to methods

2.4.3 Call Graph

This method establishes the potential effects of system changes on already existing source codes through identifying call relationships of both functions and procedures/methods that build the code. Hence, specific calls of procedures and functions should be derived from source codes and are stored in a graphical representation called a "call graph". With the completion of the calls extraction process, a call graph's transitive closure is evaluated and used for further Change Impact Analysis tasks.

In this kind of Change Impact Analysis dependency analysis, methods are determined as affected if the methods call a modified method. In other dimensions, methods are often impacted when a modified method is called by other called methods repeatedly. Related instances are depicted in [25,26]. The advantages and drawbacks of the call graph analysis approach are represented in Table 2.3.

2.4.4 DETA Analysis

Call graph Change Impact Analysis techniques suffer from extremely low accuracy metrics [27]. This can be enhanced by considering the various procedures which are called during the software program implementation. Therefore, a proposal is made to monitor the execution process of software programs for Change Impact Analysis purposes. However, this proposed method needs specified software programs to be effectively adapted and implemented to determine respective call behaviours. Finally, the execution traces which are monitored can be used for Change Impact Analysis, where any tracing model consisting of a minimum of the single modified model is regarded as affected. Related instances regarding this approach are depicted in [27–29]. The advantages and drawbacks of the DETA approach are represented in Table 2.4.

2.4.5 Program Slicing Analysis

The approach is utilized for understanding the project and maintenance of tasks. In addition, this is often employed in the Change Impact Analysis procedure. Slicing "brushes away" any such code sentences which have no impact on the specific code parameter. This results in all those code sentences which are impacted by modifications to the particular parameter.

A slice is measured by analyzing both dependencies including data (such as value assignments) and control (like if-conditions) dependencies of software program statements, which alter a program variable. Therefore, it is applicable in both the forward and backward approaches [30]. Hence, when modifying a specified statement of a software program, the slicing technique is employed to identify statements which are highly impacted by the change. Diversified program slicing tools often exist for various programming languages and Integrated Development Environments (IDEs) like those mentioned in [31] or [32]. The advantages and drawbacks of the program slicing approach are represented in Table 2.5.

2.5 MSR (MINING SOFTWARE REPOSITORIES) TECHNIQUES

Unlike other proposed techniques, MSR approaches do not involve retrieving dependencies from artefacts. Instead, these methods derive

TABLE 2.4

Benefits and Drawbacks of the DETA Approach

Benefits	Drawbacks
More precise than call graphs	Only relevant on source code
	No distinction among diverse changes
	Requires code instrumentation
	Granularity restricted to approaches

TABLE 2.5

Benefits and Drawbacks of the Program Slicing Approach

Benefits	Drawbacks
Acceptance by programmers	Only relevant on source code
Good tool support	

artefacts from repositories of software. While maintaining track of the version history of artefacts, this approach examines efficient associations between various software artefacts which are only accessible from repositories. Few of the broadly used version control systems include CVS, SVN or Git.

The MSR-based analysis is engaged in exploring the patterns of co-modification or co-evolution of various software artefacts if there exist any regular alterations between those artefacts. Thus, by analyzing the probability of frequent co-changes, MSR methods will be able to identify the correlated artefacts. Instances of this approach are explained in [23,33–36]. The advantages and drawbacks of the MSR-based approach are represented in Table 2.6.

2.6 INFORMATION RETRIEVAL (IR)

IR-based techniques are used for name scanning, detectors and text modules of software artefacts to extract the patterns with a similar text. If any textual matches between two software artefacts are observed, the IR approach estimates that change in one artefact will obviously affect the other. As in MSR methods, IR-based techniques are effectively involved in analyzing various types of artefacts, considering the

TABLE 2.6

Benefits and Drawbacks of the MSR-Based Approach

Benefits	Drawbacks
Multi-perspective analysis possible	Requires history availability
	Artefacts should evolve in a similar repository
	Dependent on commit behaviour
	Selection of sliding window critical
	No distinction among diverse changes
	Not applicable in the early stages

TABLE 2.7

Benefits and Drawbacks of the IR-Based Approach

Benefits	Drawbacks
Multi-perspective analysis possible	No distinction among diverse modifications
	Only lexical similarities deliberated

textual components of artefacts. IR-based impact analysis methods are broadly classified into two phases: pre-processing analysis and original text analysis. On the pre-processing analysis front, essential techniques like remove-stop-word and stemming-word are utilized to minimize the possible space for search [37]. Later, the actual text phase is initiated by utilizing N-gram-based [38] or Latent Semantic Indexing [39] text comparison methods. However, these techniques neither describe a difference between diversified types of change activities nor identify the resulting effects. Instances related to these approaches are depicted in [40,41]. The advantages and drawbacks of the IR-based approach are represented in Table 2.7.

2.7 PROBABILISTIC APPROACHES

Some of the studies that incorporated probabilistic techniques are introduced to represent and analyze a specified software system when a change is observed in the system. These approaches utilize different probabilistic models, including Markov chains to analyze the affected system. With the completion of system modelling, the probabilistic Change Impact Analysis approach is initiated for executing Change Impact Analysis functions like Granger causality tests or Bayesian inference. Hence, as per the computed probabilities, if a change in software artefact is noted then related information is sent to the software developer. Instances related to this research work are explained in [36,42,43]. The advantages and drawbacks of the probabilistic approaches are represented in Table 2.8.

2.8 RULE-BASED TECHNIQUES

A unique Change Impact Analysis procedure is proposed through rule-based methods. According to this approach, a definite set of rules are used to predict the various effects of a system change.

TABLE 2.8

Benefits and Drawbacks of Probabilistic Approaches

Benefits	Drawbacks
Multi-perspective analysis possible	Difficult to understand the computed change propagation
	Hard to model the interplay of heterogeneous software artefacts
	No distinction among diverse changes

The primary objective of this method is to detect an impact through established rules when specific modification activity is carried out in a given software system.

The process of establishing an essential set of rules varies from method to method. Few researchers advise using the information of a developer, domain and design methodologies to generate the most appropriate Change Impact Analysis rules. With the dependency on software artefacts types, multiple query languages are employed to execute these set of rules, including those given in [44–46]. The case studies which relate to this approach are depicted in [47,48]. The advantages and drawbacks of the rule-based approach are represented in Table 2.9.

2.9 HYBRID APPROACHES

Hybrid approaches include a combination of numerous above-mentioned techniques to execute an efficient Change Impact Analysis. For instance, in [49], the researcher makes use of diversified analyses

TABLE 2.9

Benefits and Drawbacks of the Rule-Based Approach

Benefits	Drawbacks
Multi-perspective analysis possible	Rules creation is time consuming
Can be enhanced with repair plans	Rules creation is not addressed by the present research
Can address different changes	Rules need maintenance
Allows addition/change of rules	Hard to address ambiguous impacts
	Rules require validation

including IR approaches, MSR techniques and DETA approach. Despite being involved in identifying the significant effects of a system change in different artefacts of software, this approach is unable to classify the accurate impact types. In addition, as it is combined with other methods, it also experiences the drawbacks of those approaches.

2.10 MULTI-PERSPECTIVE APPROACHES

One of the main challenges of the review of literature is to check whether there are any methods present for the multi-perspective impact of change analysis that are meeting our pre-requisites. Therefore, in this segment, we represent outcomes of review with respect to perspectives and software artefact types covered by contemporary researches. As per our taxonomy of impact analysis, here, we deliberate three important software artefact types and their relevant perspectives like architectures of software, pre-requisites and source code. Further, some contributions also examine documentation files and configuration that are referred to as other artefacts because of inhomogeneous reasons and architecture. Figures 2.1 and 2.2 summarize the results of our literature review.

Also, we examined how many distinct software artefacts and perspectives were covered by some of the current multi-perspective methods. Moreover, distinct perspective distribution summarization is shown in Figure 2.2, among some of the available multi-perspective methods.

DISTRIBUTION OF LITERATURE SCOPE

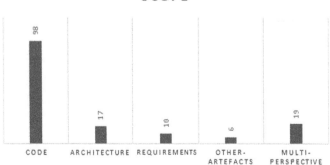

Figure 2.1 Distribution of scopes among the studied approaches.

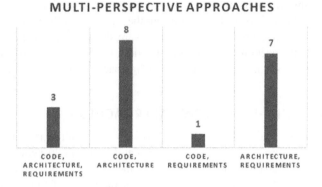

MULTI-PERSPECTIVE APPROACHES

Figure 2.2 Scopes supported by multi-perspective approaches.

The huge majority of current multi-perspective methods are concentrated in two perspectives exactly, where only three methods are truly capable of addressing heterogeneous software artefacts types (Figure 2.2). Further, most of the methods are confined strictly in their assistance of distinct kinds of varying operations.

3

CHANGE IMPACT ANALYSIS IN SOFTWARE VERSIONING
Issues and Challenges

Most software modifications involve massive complexity as the effect of such modifications cannot be accurately predicted in advance. However, with shifts in consumption patterns, evolving technologies force frequent alterations in the code. Instead of these changes, the need for an accurate evaluation of the impact of such modifications over the current functioning is strongly observed. Change Impact Analysis is described as the detection of the possible impact due to the code modification on the single section or more sections of an application. In the case of safety-critical domains with large applications, code modifications can expose the code to attacks, causing a system bug down. In such an environment, the pace of accurately predicting the impact on individual components or the whole system plays a vital role. Detecting the components susceptible to changes remains the key functioning of a Change Impact Analysis.

3.1 AREAS OF CHANGE IMPACT ANALYSIS IMPLEMENTATION

When any modification request is initiated, the change monitoring board allocates an evaluator to analyse the requested modification. The evaluator evaluates the importance of the proposed modification in the business aspect. Business Impact Analysis (BIA) (see Figure 3.1) is initially performed to ensure whether the proposed modification can add additional business value or enhance efficiency

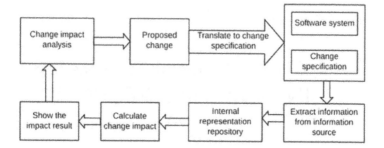

Figure 3.1 Change Impact Analysis process.

or project quality or a combination of these outcomes. In the next stage, the evaluator conducts Change Impact Analysis to detect the ripple impact of the proposed modification [18].

In addition, Change Impact Analysis is used to detect change factors, areas highly impacted by the modification, which is often based on the knowledge gained from the history of requirements. It also observes the outcomes of Change Impact Analysis and extends the analysis to further requirements including safety, size, economics, adaptability and robustness, along with human factors. Considering all the derived factors and their possible impacts, decision over implementing the change or denying it is made.

Alternatively, Change Impact Analysis is implemented when the evaluator attempts to understand the ripple effects and cost analysis of the proposed change. This process enables the strategy formulation team to understand delivery times and to promote possible additions in the releases over the stipulated time period [50].

Analysing different models and primary software modification activities, we can conclude that they are focused on reviewing programs before and after the proposed change, incorporating modifications within the available framework and ensuring proper functioning of the system after the change implementation. All these functions of the change monitoring board enable it to understand the project in the context of change [51]. Further, as [52] suggests, the change implementation process should estimate the ripple effects of change and ensure nothing is overlooked.

Following the successful incorporation of the proposed change, rigorous testing has to be performed. Test cases which require further examination or cases requiring design changes are identified in the process [52].

3.2 THE APPROACH OF CHANGE IMPACT ANALYSIS

This method has many benefits. Primarily, this method assists the change control board (CCB) and managers and is projected for detecting the impacted artefact components. It comprises associations amid components in software procedure and detects impacted components in the path at the time of traverse. Moreover, it could be performed automatically, enabling project managers to lessen the effort in detecting change or varying impacts.

When a change request is submitted by an originator towards an ongoing project, the control board of change allocates an assessor for variation. Here, the assessor performs Impact Analysis of business for detecting change value for the business and the project. Further, the assessor also verifies whether the projected request for change is completed formally and understood; this stage might occur within less time. Here, the assessor might then detect the change value towards business in the following ways:

- Whether the projected change could be assisted by the case of business or by regulatory or legal pre-requisite.
- Which consumer pre-requisite improvement chapters, sub-areas of business and performance might be developed and enhanced.
- Whether the projected change makes issues with current pre-requisites.
- Whether there were any conflicting change requests submitted already.
- Whether the requested change forms any gridlocks or bottle-necks of business.
- Whether the requested change could be deemed expensive, intricate, controversial or confusing.

Even though the assessor might always decide before consulting with CCB in an instance of a positive solution to the query, the requested change might be submitted to attain the objectives of the project in a more effective way and might enable faster cooperation. Figure 3.2 signifies the procedure of Impact Analysis of business.

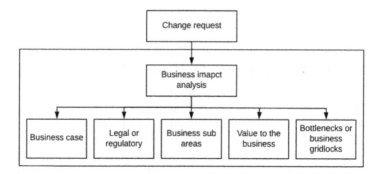

Figure 3.2 Business Impact Analysis process.

After completing the Impact Analysis of business, the assessor performs Impact Analysis of change for specifying the elements influenced by the change request. Here, in this phase, the assessor is required to:

- Signify what has been devised primarily in the application of the project: cost item name or activity, budget line and present situation and a package of work.
- Detect a suitable method for variation.
- Convert the projected change into the specification of the system change.
- Extract the data from source information and transfer it into the internal repository of representation.
- Improve the predictions of resources on the basis of deliberations like intricacy of software and size.
- Assign the resources for reviewing the change request impact.
- Initiate the activity, which is direct for evaluating the price and scope and scheduling the change influence.
- Update requested change with the analysis of impact and predict scheduled price, effort and scope.
- Describe the changes of request and justify them for variation or change: describe the pre-requisites of changes and purpose, and signify from where the changes of sources would be recovered, if associated, by signifying the amount of line of the budget, price item and work package.
- Describe the requested change's impact on the application of the project.

3.3 CRITICAL ISSUES OF CHANGE IMPACT ANALYSIS

A significant limitation of Change Impact Analysis observed in recent years is the data source volume. Despite the Change Impact Analysis being implemented in multiple forms by several studies, no standard definition exists. For instance, IEEE has no mention of Change Impact Analysis in its glossary and regards it as a concept-based analysis.

Furthermore, ripple effect and predicting modification and its effectiveness of Change Impact Analysis implementation remain tough challenges. In addition, identifying the suitable method of Change Impact Analysis remains the most dangerous stage of the entire implementation process. Other problems in executing Change Impact Analysis include the need to obtain and organize all the possible consequences of change and manage these consequences appropriately [17]. On the other hand, in small decisions, Change Impact Analysis can be implemented as a simple desk work [53], and these organizations rarely attempt to understand Change Impact Analysis.

Apart from these limitations, unavailability of dimensions for evaluating Change Impact Analysis approach performance with other approaches is also one of the shortcomings. Further, it is difficult to understand if the existing information is adequate to compare Change Impact Analysis approach [54]. Accordingly, it remains challenging to detect those works contributing to Change Impact Analysis. Further, identifying the most suitable analysis approach for the specific software is related to software modifications.

Observing modification consequences by relationship analysis is also a complex task. Two strategies are used for Change Impact Analysis: inductive and deductive. The inductive approach iteratively navigates program life-cycle object relation data from storage and detects parameters with considerable effects. On the other hand, the deductive strategy is engaged in deciding if the life-cycle object has been impacted by the modification [16].

3.4 ROLE OF RELATIONSHIPS IN CHANGE IMPACT ANALYSIS

Change Impact Analysis remains the essential technique for handling modifications. The concept aims to detect the relationships among components. On the basis of dependence or independence, Change

Impact Analysis attempts to observe the effect of the proposed change to the remaining artefacts. As discussed earlier, the requirement framework comprises artefacts and associated relationships between these artefacts. Typically, the relationships in Change Impact Analysis are categorized into four kinds – aggregation relationship, composition relationship, necessary impact and supplementary impact.

Of the four types of relationships, aggregation refers to an artefact within another artefact. The aggregation relationship depicts the time period in which one artefact is independent of the entire artefact [55].

Impact relationships depict if the source artefact is a compulsory feeding parameter to generate the target artefact. Existing impact relationship value entails that the target is compulsorily impacted to any modifications made to the source [55].

Supplementary impact relationships depict if the source acts as a supplementary parameter for generating the target. Existing relationship value implies that any changes in the source may not impact the target. Accordingly, these relationships enable us to understand the effect of proposed modification accurately so as to decide whether to implement the modification or not.

3.5 FACTORS TO CONSIDER FOR EVALUATING CHANGE IMPACT ANALYSIS IMPACT

3.5.1 BIA Analysis

- Verify the suggested modification fits into business strategy and/or meets regulatory norms
- Identify sections that will be impacted due to the proposed modification
- Verify if the proposed change is compatible with the existing framework
- Check if any conflicting modifications already exist
- Assess possible modification risks and potential roadblocks
- Evaluate the extent of costs involved and the complexity of the modification and check if any controversies exist

3.5.2 Change Impact Analysis

- Pre-determine the project name, costs involved, budgets and current conditions
- Determine the acceptable procedure for change implementation

- Delve down on the suggested modification into system modification specifications
- Obtain data from the source code and modify it into the internal storage
- Plan resource requirements on the basis of size, complexity and time involved
- Allocate required resources to evaluate Change Impact Analysis impact
- Redesign the modification request based on estimates of scope, overheads and time period
- Obtain details on the needed modifications and sources being assigned
- Analyse Change Impact Analysis impact on implementation

3.6 SUMMARY

As discussed in the section 3, Change Impact Analysis involves iterative sequences. In particular, handling modifications is based on modifying decisions on the basis of requirements. Modification decision varies with outcomes of Change Impact Analysis. As the factors affecting the change are interrelated, any modifications in the project re-impact these interlinked factors.

4

CHANGE IMPACT
ANALYSIS METHODS
The Review of Contemporary Literature

4.1 OVERVIEW

Impact Analysis estimates the various sections of a program, which can be affected due to system modification. The status of class artefacts is essential to choose an appropriate technique to conduct Change Impact Analysis in the improvement and maintenance segment, though the static Change Impact Analysis method does not need completely developed class artefacts. This kind of approach aims to over-estimate the quantity of impacted classes in the program. Accordingly, the approach has limited utility in practice. In contrast, the dynamic approach estimates completely developed only class artefacts. This study aims to predict only if all these artefacts are undeveloped.

This chapter describes the capabilities of the current Change Impact Analysis approaches with significant intention to aid the execution process of the development phase. Based on this analysis, the primary requirements to evaluate a novel Change Impact Analysis method, particularly for the improvement phase, are then developed.

4.2 CONTEMPORARY CHANGE IMPACT
ANALYSIS APPROACHES

Change Impact Analysis techniques are broadly classified into two categories [56]: static and dynamic Change Impact Analysis approaches. The static Change Impact Analysis techniques approach is engaged in developing a group of possible affected classes through evaluating the project's static information that is developed from different artefacts. A few of software artefacts include requirement artefacts, design artefacts, class artefacts and test artefacts. In contrast, the dynamic approach is involved in developing a group of possible affected classes by assessing the dynamic information. The following sub-section describes each Change Impact Analysis category.

4.2.1 Static Analysis

The core techniques of static analysis which are employed for the novel proposed approach include [3,12,57]:

A. *The UCM Static Technique* [3]
This method employs the UCM (use case maps) method [58] to operate Change Impact Analysis on both functional requirements and the design method. A UCM technique estimates that all potential functional needs and design methods are fully developed.

UCM technique comprises two limitations: (1) it lacks tracing approaches utilized from functional needs and design artefacts to the original code. Such approach presumes that the content of both artefacts depicted through the UCM model is represented in the artefacts. The impacted artefacts indirectly represent all impacted components of the model. (2) This technique does not include either dynamic or source code analysis. However, a few consequences of a system change from one class to other are detected only by analyzing the behaviour of the modified class using behavioural analysis. Depending on this rule, this technique leads to the missing of a few affected classes.

B. *Need Interdependency Approach* [57]
Both need interdependency method and horizontal tracing method are employed in this technique. The need interdependency approach uses the methods of need interdependency identification [59]. On the other hand, the horizontal tracing approach analysis employs the method of IR [60]. As such [3], this interdependency technique estimates that all the class artefacts and requirements are fully developed before executing the method.

The major benefit of this approach over the UCM is that it detects the relationships between requirement and class artefacts by using traceability methods. This detection procedure is significant because of the specific impacts caused by changes that efficiently traverse to the class artefacts. Like UCM, it has two limitations: (1) firstly, it excludes design artefacts. It's a fact that few requirements can map with the class artefacts directly [61]. In contrast, few requirements utilize high-level design artefacts as a negotiator to reflect the artefacts. Hence, detecting all affected classes using impacted requirements is not always possible because few cases need design decision to effectively aid the detection process [13].

(2) Similar to UCM, this technique lacks both dynamic and source code approaches. However, a few consequences of a system change from one class to another are detected only through analyzing the behaviour of the modified class [29] using behavioural analysis. Depending on this rule, this technique leads to the missing of a few actually affected classes.

C. *Estimating Class Interactions through Impact Prediction Filters (CIP-IPF) Approach* [12]

CIP-IPF technique implements a class interaction estimation approach to identify the affected classes. The implementation process of CIP-IPF is almost the same as the need interdependency method [7], as modifications to a requirement reflect corresponding artefacts. However, the factors that differentiate these techniques include: (1) it generates its own need-interaction identification model, (2) employs rule-based approaches [62] for the horizontal tracing identification and (3) estimates which classes are incompletely developed and which of them are totally developed. CIP-IPF method is strong compared to the above two methods [3,57].

This approach includes tracing path identification among the need artefacts and class artefacts and enables the ease of detecting modification assessment at the need stage to class artefacts. Later, contrary to need interdependency approach, this method induces design artefact concept in the procedure to discover those class artefacts which are affected by the change.

4.2.2 Dynamic Analysis

Two approaches are proposed in [5,6], including the influence mechanism approach and path effect approach, respectively. Both of these approaches estimate the impact set on the basis of method-level observations.

In [5], the researchers induced an IG graph to detect the classes affected by the change. The graph utilizes class artefacts as the primary basis of observation and presumes that all these artefacts are fully developed. This ensures that only these impacted classes will be dynamically analyzed for further refining to identify the most affected classes. Accordingly, superior performance can be predicted through this approach. However, the model is not without shortcomings. No formal tracing procedure exists from requirement or design to class artefacts. The procedure is vital in the total Change Impact Analysis procedure as modification occurs from both class and design/need artefacts. As different design artefacts interrelate both vertically and horizontally, modifications to these artefacts can add to various impacted class artefacts. In certain conditions, solely relying on the source code can ignore such impacted class artefacts.

4.3 EVALUATION FACTORS

For evaluating the efficiency of different Change Impact Analysis approaches concerning their assistance in the program development stage, this study designed four factors of review or evaluation [3, 4].

These factors are chosen because of their role in assisting Impact Analysis in the development stage. The four factors include: (1) information source, (2) model design approach, (3) partly built class consideration and (4) analysis.

Information source factor: This factor plays a key role in building the Change Impact Analysis model. A few approaches rely on the class artefact, while a few of the strategies rely on either need artefact or design artefact as the information source. In strategies for the program development stage, the need and design artefacts are regarded as highly stable and developed and are most preferred in practice [3,4].

Model design approach factor: This factor depicts the approach incorporated for impact determination. Two approaches are employed for building the model – forward engineering and reverse engineering. As the class artefacts have strong limitations for being used as the information source for model building, the reverse engineering approach is not regarded as a practical model design approach [6].

Partly built class consideration factor: This factor determines the efficiency of any Change Impact Analysis approach to incorporate partly built class analysis. It plays a vital role in the analytical process as only specific classes in the improvement stage are completely developed [63].

Analysis factor: This factor establishes the role of an approach analysis in detecting the affected artefacts. This analysis is conducted in two ways – either static or dynamic. A few approaches incorporate the static method, and a few opt for the dynamic method. Specific approaches include a combination of both of these methods.

The hybrid technique combining both of these methods mostly prefer static methods as they have a strong advantage in cases of partly developed classes, and the dynamic analysis results in superior performance only in the case of fully developed classes. Nevertheless, dynamic methods are incorporated only if certain potential impacted classes identified in the static method are completely built.

4.4 ANALYSIS RESULTS

The following description denotes the existing approaches of Change Impact Analysis by comprising the factors of (i) critical objective of the analysis, (ii) analysis model, (iii) considering the incomplete classes and (iv) analysis mode.

Table 4.1 explains that the existing techniques of Change Impact Analysis are unable to support each significant element to execute the Change Impact Analysis technique in the improvement phase. For instance, the UCM techniques only aid the analysis element source by utilizing high-level software artefacts, whereas the model building approach factor is based on the estimation method. This approach, however, excludes partly developed classes and relies solely on static analysis. Accordingly, this approach cannot be efficiently implemented in the development stage. Therefore, we observe that the novel Change Impact Analysis methods are to be developed.

The novel Change Impact Analysis technique will demonstrate four significant characteristics:

- Utilizes high-level software artefacts (for instance, requirement artefacts and design artefacts) as the resource to generate a Change Impact Analysis approach
- Applies appropriate predictive techniques as the Change Impact Analysis method building techniques
- Comprises incompletely developed classes so as to execute dynamic Impact Analysis
- Merges both static and dynamic techniques in order to execute efficient Impact Analysis

TABLE 4.1
Current Impact Analysis Techniques' Strengths and Weaknesses

Method	Objective	Model	Analysis of Incomplete Classes	Analysis Mode
Prediction of interconnected classes	Analysis of application design and requirement artefacts	Predictive analysis	NO	Static
Assessment of coverage impact	Analysis of either source code or class-level artefacts	Reverse engineering	NO	Dynamic
Defining influence graph	Analysis of either source code or class-level artefacts	Reverse engineering	NO	Static and dynamic
Predicting path impact	Analysis of either source code or class-level artefacts	Reverse engineering	NO	Dynamic
Mapping interdependent requirements	Analysis of requirement artefacts	Predictive analysis	YES	Static
Mapping use-case diagrams	Analysis of application design and requirement artefacts	Predictive analysis	NO	Static

4.5 SUMMARY

This chapter represents a complete study of the existing methods of Change Impact Analysis, predominantly to aid the development phase of software execution. From the above work, we state that the novel Change Impact Analysis technique has to comprise four significant characteristics as depicted above. To support future research work, we focus on developing a new method of Change Impact Analysis depending on the four aforesaid features.

5

CHANGE IMPACT ANALYSIS IN MAINTENANCE PROJECTS

5.1 OVERVIEW

The increasing complexity of software applications is leading to the increased complexity involved in the project design. In addition to code maintenance works, build configuration maintenance works are gaining prominence to ensure that unintended disruptions do not occur and slow down system functioning [64]. The researchers in [65] observed that poor build maintenance efforts result in the breakdown of the build configuration. The study in [66] reiterated the importance of build maintenance by studying the costs occurring due to building breakages. Any modifications in source or test codes also cause changes in build configuration files. In [67], Machine Learning techniques are proposed for training models to estimate build modifications on the basis of earlier changes in both sources and build files. The study in [68] further emphasized the feasibility of making these estimations across multiple projects.

This chapter attempts to improve the performance of these methods by modifying the source code and commit categories. The approaches suggested in [69,70] are utilized to extract source file modification features. A Change Distiller is a plugin that obtains code modifications on the basis of tree-differencing [71]. It extracts two versions of the source code and generates respective ASTs (Abstract Syntax Trees). By comparing these two ASTs and mapping the differences to respective code modification classes, Change Distiller obtains changes. To classify commit messages, [72] suggested the division of these messages into four classes according to the content.

5.2 CHANGE IMPACT ANALYSIS TECHNIQUES FOR SOURCE CODE MODIFICATIONS

The following paragraphs depict multiple Change Impact Analysis techniques suggested in different studies. The approaches which were empirically verified are presented.

Most Change Impact Analysis approaches rely on conventional static analysis models based on the dependency map. For instance, an impact set is calculated based on reachability as depicted in a dependency map. To boost the accuracy of Change Impact Analysis approaches, a few studies suggested [27–29] Change Impact Analysis approaches on the basis of data obtained in the code execution phase. Some of such information can be trace data, coverage or associated data, used to create the impact set.

In addition, specific Change Impact Analysis approaches [35,40] function on the basis of coupling metrics like a structural or associated topic or dynamic working. The related impact sets are estimated based on code sections coupled with modifications.

Further, the studies [5,27,36] focused on the extension of existing Change Impact Analysis approaches. Of these, two studies [5,27] were extensions to other contemporary methods [27,36] with minimal time and space overheads than the original. The study outcome suggested almost the same accuracy for both online and original approaches. The study depicted the superior performance of fine-tuned Change Impact Analysis at code level regarding accuracy compared to the original approach. However, the extended model involved more complexity and thereby more time periods were required. The researchers in [36] compared hybrid techniques by mixing the ranking of association rules with Granger [43].

Analyzing different Change Impact Analysis approaches, we can observe that almost all of these fall into four categories – conventional dependency based, mining from storage based, coupling metrics based and execution data based. Evolution of these studies over multiple years is depicted in Figure 5.1. As can be observed from the figure, mining approaches are the most focused studies with increasing interest in the past few years. Mining from repositories enables Change Impact Analysis analysts to extract valuable dependency data between different artefacts. Maintenance engineers rely on this historical information to estimate modifications to associated artefacts. On the contrary, simply depending on conventional code dependencies can have the possibility of not obtaining key dependency data.

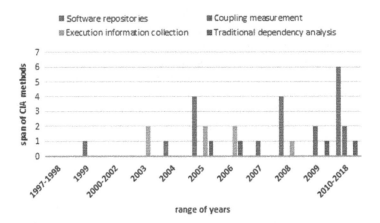

Figure 5.1 Distribution of the Change Impact Analysis techniques from four perspectives.

For instance, any code modifications recorded in a file may need related modifications in the code to enable reading such data irrespective of whether the data flow exists between the two code sections. Accordingly, stored data emerges as vital support for Change Impact Analysis approaches based on conventional methods.

In addition to mining-based studies, dynamic Change Impact Analysis approaches are also studied largely in the recent period. This is largely due to their ability to enhance the accuracy rates of impact results compared to conventional static approaches.

Predicting modifications – several research studies can be found in contemporary literature, focusing on estimation tasks, specifically deploying ML algorithms to analyze different features of programming. The study in [73] evaluated if a programmer must participate in email communications, while the study in [74] predicted fault proneness of the source code. In [75], researchers detected modification-prone language interfaces, and in [76], future bugs are predicted.

5.2.1 Change Impact Analysis for Software Framework

Most of the Change Impact Analysis techniques suggested in the literature have relied mostly on source code files [77]. This resulted in not many studies focusing on different artefacts of the program.

However, in the recent past, researchers have been working on extending Change Impact Analysis for the framework and design tasks. The study in [78] and [42] focused on the understanding of modifications in the source code and UML design. It used the conventional dependency approach for establishing the impact of adding or removing options. Sharafat and Tahvildari used probability-based approaches for determining the same. However, given the similarity of both source and UML files, they fail to offer solutions for understanding other UML methods like component maps. Further, they also restrict the analysis of the source code. Accordingly, these techniques can be regarded as a single-perspective and not as a multi-perspective.

In addition, the studies in [79] and [80] observed the proliferation of modification among different framework segments and system requirements through dependency techniques. This remains a vital phase for understanding the total proliferation and connectivity to the different framework, and implementation artefacts are needed. This ensures the complete prediction of assessing modification effects. However, a large gap exists between framework components and design artefacts.

A detailed analysis of existing studies suggests that two techniques can be implemented for handling heterogeneous program artefacts. However, even these techniques have several limitations, leading to further scope of work in the area. The following paragraphs detail the shortcomings of these techniques.

The study in [47] suggested a Change Impact Analysis technique and regression analysis for evaluating test and use data, class and sequence diagrams. The researchers suggested a new technique for categorizing test data based on its ability to be re-used, further testing or outdated status. The proposed Change Impact Analysis technique can be categorized under dependency Change Impact Analysis. The method could not include key UML artefacts like frequently utilized components. It also does not cover activity diagrams, restricting the use of the technique in practice and its extension to assess multi-perspective Change Impact Analysis. Further, the study also ignores artefacts included in the source code.

The techniques studied in [81,82] focused on dependency approaches for observing test data and classes, among others. Dependencies are extracted from three sources. From the static source, the call graph is obtained; test cases assist in obtaining interdependency among models and needs. Mining techniques are employed to obtain more dependencies. Information from these three approaches is saved as traceability links. These links enable researchers to establish a proliferation of modifications among different artefacts. Similar to earlier studies, this study also fails to address multi-perspective modification requests. It also does not handle all source artefacts or framework components. It also needs instrumentation and involves human interference, resulting in large overheads for implementing the technique over a reasonably complex application.

6

CHANGE REQUEST IMPACT ANALYSIS TOOLS

6.1 OVERVIEW

Several researchers have identified that performing the analysis of the impact at the time of maintenance and improvement stages of the life cycle of software has many advantages [83,84]. Several tools are developed for researching, which summarize the Impact Analysis merits. However, there could be a lack of noticeable tools made accessible to commercial developers.

The objective of this chapter is to research distinct tools of Impact Analysis, which are present either commercially or academically, and classify them per improved architecture. The other aim is to detect the tool's features, which are resourceful for the developers to execute maintenance. The consequence of the findings needs to offer a base for further enhancement of the tool of Impact Analysis, which developers might utilize for code maintenance and evolution.

With increasing popularity, incessant combining and scrum or agile development, the rapid, effective evolution of code is becoming more significant. The code is committed by the developers, and consequently, a server is built which tests the builds and analyzes the static code. In incessant integration, while each developer is committing towards a similar baseline, it could be enormously crucial that the developer evades bad code committing whenever possible. The poor committing might turn to damage the build that stops any further checking of the analysis of the static code or tests from running.

The responsibility of developers is to assure that their commitment would not damage the build; however, it might still occur and could possibly disturb the workflow of the team.

Locally, running the regression tests is one of the models which developers utilize for assuring that their commitment would not damage the build. The developers select to run total tests or choose manually the subset of the tests, which are visible to be associated to set change. One of the confines is that locally running the test might take more amount of time, specifically when the developer selects to run entire tests at a time. Besides, the developer also selects incorrect tests subset that might still result in damaging the build.

Moreover, there were several strategies for Impact Analysis which were explored and developed. However, some of the tools were perceived to offer these strategies for developers of software. These strategies have exhibited the ability to enhance the maintenance of software and lessen the quantity of the error established into the significant baseline. It stands for a reason, where these strategies need to be condensed into tools usable by developers. They also identified the requirement for the effective tool development of Impact Analysis to assure continuous utilization and widespread strategies related to Impact Analysis [85,86].

The objective of the effective maintenance of software is to provide developers tools for detecting the propagation extent of changes and enable the developer to make a decision as to whether it is secure for committing the code. By analyzing and cataloguing the abilities offered by the current tools of Impact Analysis, we introduce what is presented today and detect any possible scope or further tools of Impact Analysis.

There are several tools which are present to execute the analysis of the impact, and only the subsets of tools have been selected for analyzing. This contribution concentrates on tools which are aimed by the developer in the development of the pre-commit stage before modifications have been combined into the main baseline of the project. This chapter discovers the tools which assist developers in answering for the query: What might happen when I make this modification? And they need to possess a degree of confidence regarding the integrity of an outgoing set of changes.

It could be recorded that there were distinct tools available which aid the realm of maintenance of software for the analysis of the impact, like the analysis tools of the static code. The analysis tools of the static code like check style 2 or find bugs 1 might greatly enhance the effectiveness and stability of the code by developers, by warning the developer for violating the standards of coding. These tools need to be utilized in combination with the tools of Impact Analysis to assure that the code has been combined in greater quality and would not impact the current baseline negatively.

Moreover, there are also an number of tools which assist the developer in understanding the framework and the association of the code, and most of them comprise elements of visualization. These tools are deliberated; however, most of them were omitted when they could not implement an intelligent model or algorithm, including the analysis of the impact.

It could also be recorded that the analysis of the impact term could be an extensive term. The analysis of the impact might be performed at several levels from a level of code to component of a system and even at the level of conceptual. Several tools have discovered which covers these distinct analyses. However, several were signified to be out of scope. The objective of this chapter is to concentrate on an analysis of the impact at the level of code by utilizing algorithms, which are able to analyze or parse the languages of programming.

6.2 METHODS OF IMPACT ANALYSIS

6.2.1 Program Slicing

The work [87] proposed program slicing in the form of program parts isolation, which associates to a specified point of a program. The specified slice of the program could be a program executable and isolated segment. With the analysis of the slice of program, the developer might understand associations among statements and variables that might aid in the optimization and debugging of the program.

The work [88] researched former contribution performed by utilizing methods of program slicing and explored and compared every method. Two of the methods examined are dynamic and static slicing that are proposed. Besides, some of the distinguished slicing-associated vocabulary is in the following way:

- Static slicing: Utilizing slicing of a program as per criteria; however, it is not dependent on the execution of a program.
- Dynamic slicing: Utilizing the slicing of a program with a particular performance for analyzing the conduct of the program.
- Backward slicing: Detecting statements which might impact the criteria of slicing potentially.
- Forward slicing: Detecting the statements which are impacted by the criteria of slicing potentially.
- Associated slicing: Beginning with active slice, isolating statements which generally impact the criteria of slicing.
- Decomposition slicing: Detecting the program statements which are required for computing a specified variable.

The slicing of a program might also offer input into metrics of coupling and cohesion. Coupling could be the mutual interdependence degree among modules of the software. Typically, when it could be computed by information measuring in and out of specified modules, the work [89] identified that utilizing the slicing of the program leads to a more accurate computation. The work [90] also examined the coupling measurement utilization in the form of ripple impact, discovering the entire object-oriented program.

The work [91] presents that in addition to the measurement of cohesion and coupling, there were several other uses of slicing or a program like testing, integration of program, parallelization, reverse engineering and refactoring. The work [91] presents the research of distinct methods of program slicing and compares the effectiveness and accuracy of the models. For categorizing the methods, the features of the succeeding language of programming were utilized: control the flow of unorganized pointers/variables of composite and concurrency.

6.2.2 Impact Analysis and Repository Mining

The work [92] presents that a model is created to mine the repositories of software for facilitating the analysis of the impact. The strategies for the retrieval of data and ML were utilized along with data from the Mylyn for capturing information regarding entities which were communicated previously. This method could be detected for resulting in optimum recall gains when compared with other mining methods of Software Change Management (SCM).

6.3 DESCRIPTIONS OF TOOL

In this segment, we might offer some background data and details on every 20 analysis tools of the impact under the observation. When we have developed an architecture for categorizing every tool as per common traits and properties, every tool could be unique in its own way.

Unravel: The work [93] presents that unravel could be an open source tool for program slicing of count 13, which is available for C programs by NIST. The objective for a tool is to assist developers with the comprehension of a program and debugging by detecting the slices of the program with specified criteria of slicing. Moreover, the tool is assessed to define when the produced slices were of resourceful size for the programmers, when the tool calculates the slices efficiently and quickly and when the tool was deliberated for use by an average programmer.

This tool is formed by three parts: slicer, linker and analyzer. It initially analyses the source code and slices it into independent modules, and these modules are mapped to corresponding source and configuration files.

SLICE: The work [94] presents that SLICE could be an active slicer for programs of C. The tool assists both forward and backward slicing and also delivers four types of slices for aiding the developer in the procedure of debugging: closure of data, executable, reliance of data and expected results or outcomes of the executable. For performing active analysis, the specified insight is implemented for the program to get executed, and slices are built for any related variables as per the criteria of slicing. In the performance of SLICE assessment, it is confirmed that the outcome of active slices was small when compared to the program size. The sizes of slice might change from program to program; however, this might be impacted by the selected criteria of slicing.

The SLICE delivers two modes for a developer: the communicative mode, where a developer might select the variable for slicing and notice the outcomes exhibited, and the mode of a batch, where operations of slicing could be performed on manifold programs by outcomes summary.

CodeSurfer: The study in [95] presents that CodeSurfer is developed originally as a tool of research at Wisconsin University, and it was packaged eventually into commercial equipment delivered by the grammatech 2. The tool delivers extensive features ranging from assisting developers in understanding C++ and C codes better, involving the analysis of pointer, analyzing dataflow and analyzing the impact.

The API could be given to developers, for enabling them to combine the abilities of the analysis directly and make the platform of CodeSurfers to benefit from depicting the functions of AST. The grammatech might provide another equipment "codesonar" that has the functions of static analysis borrowed from the CodeSurfer. The developers are assisted by codesonar for detecting inconsistencies and bugs, and optimizations are suggested with regard to security and performance.

REST: The work in [96] presents that REST is a tool, which applies an algorithm that computes the ripple impact, where the specified change is having a program rest. The programs implemented in C assist the developers by the tools with four significant activities: define probable effects, detect impacts that are known, find software constant and verify pre-requisites.

For computing the ripple impact, the information of propagation is gathered and sorted into matrices. By linking information among matrices, algorithms could be defined on how values were propagated within distinct modules. By this data, developers might take decisions for designing safely.

Chianti: The work in [25] presents that Chianti could be a tool of Impact Analysis of change available as plugin 1 of an eclipse, which detects atomic modifications set depicting the variance among two versions of a specified program. From the changes in the set, the potentially impacted set of regression tests was detected.

The concept behind this tool is to be able to perform dynamic and static analysis; however, the dynamic analysis could be utilized for illustrating the instances. The primary stage in the model is compiling atomic changes set that are classified by the change type like added class, methods changed or fields deleted. Later, graphs are called and built for every regression test, and atomic modifications are interlinked with their counterparts. For the final stage of every regression test, the related atomic changes set could be correlated and detected.

eROSE: The work in [34] presents that eROSE could be plugin 3 of an eclipse, which combines with CVS for mining the history of version for the specified project, and the user is guided into understanding the results after making the modifications. The work [97] presents that identical to impact miner; the eROSE is having the capability of recommending changes, which might be a pre-requisite for preventing errors on the basis of former commits of control version. Recommendations are ranked through assistance and suggestion of confidence level.

For performing analysis, the eROSE utilizes the server for gathering the change of transactions from CVS, and related files were parsed. The further stage is mining policies from transactions. Recognized policies frequency could also be defined, and this could be utilized for allocating the level of confidence towards every rule. More often, the rule or pattern occurs in repository; the high confident algorithm could be that it might be a suitable recommended change.

Lattix: The work in [98] presents that Lattix has eleven commercial tools, which might be used by the developers, managers, Quality analysts and architects in understanding the framework of the project and effect of the changes in the entire life cycle of software. The solution of the enterprise system is delivered, which enables developers for performing the analysis of the impact and understanding how possible modifications might impact the system. Moreover, for assisting distinct languages beyond C++, C and java, the Lattix has the ability to combine Klocwork 12, the well-known framework of static analysis to detect probable failures.

EAT: The work in [28] presents that EAT (execute-after-tool) has proposed to assess the method called CollectEA that targets to exhibit the advantages of dynamic analysis on static analysis. The programs are written in java; the EAT is made of three elements: runtime monitors set, a module of analysis and module of instrumentation.

The method of Collect EA is exhibited to be accurately similar to other dynamic analysis methods and could also be time effective. The method granularity only moves down towards the level of the method; however, enhancements could be done for capturing modifications down the level of a statement.

Indus: The work in [99] presents that Indus could be the slicer tool of java program of open source 6, which provides analysis models for assisting developers in emerging programs of java. There were three significant functionalities of the tool: slicer of java program, a gathering of static analysis abilities, and correlating with sources.

The tool offers a UI, which enables developers to specify the parameters of slicing like a type of slice or searching for a slice that is executable. When there is an availability of GUI, the Indus could be meant for being utilized as a library. The work in [100] presents that Kaveri could be one of the recognized tools, which applies to the API of Indus.

JRipples: The work in [101] presents that JRipples could be plugin 8 of an eclipse, which aimed to assist the developers by change propagation tasks and incremental modification. With three elements, it examines dependencies within the project of java, result set, and impacted classes. Then, results are reviewed by the developer individually and are marked to be visited or impacted.

JRipples applies the integration of search method dependency and retrieval of information, which is signified as DepIR for performing the analysis impact. This could be meant to mimic the actions of a programmer performing the analysis of the impact. When an entire analysis has been completed, the outcomes are exhibited in either a table or hierarchical view. From here, the developer might iterate the modifications for determining what has to be fixed.

JArchitect: The work in [102] presents that JArchitect could be commercial 7 tools of the static analysis of java programs. It delivers an extensive range of features, involving metrics of code quality, monitoring the trend, abilities to diagram and a graph of interactive dependence. The developers might elect the entity in the project, and JArchitect might show the graph dependency that is impacting fields, methods, packages and projects. The other graph types were capable of understanding associations among elements like path, cycle and coupling graphs.

For performing the Impact Analysis, the developers might view the dependency generated graphs for getting the idea of programs association for making informed decisions regarding the propagation of change. The JArchitect also produces a matrix of dependency structure that could be utilized for collecting data regarding coupling among entities. It needs to be recorded that the analysis of impact performance offers down granularity towards a level of method. The JArchitect could be accessible under two distinct commercial license types. Besides, developers have an option for choosing a free trial period of 14 days.

Kaveri: The work in [100] presents that Kaveri could be a tool of program slicing provided as plugin 10 of an eclipse. The work in [99] presents that it could be built by Indus, the slicing library of java program that might perform forward or backward slicing. Kaveri slices by abstracting unimportant details to optimize the procedure of analysis. The criteria of slice could be selected, and consequent slices could be highlighted in the editor of eclipse.

The developers might use Kaveri for tracing the dependencies of the program in augmenting the general comprehension of the program, prepared to modify propagation or detect the possible error source. Slice could be displayed in the editor of eclipse, and the developer has an option for performing slicing additive or slices intersecting on the basis of manifold criteria.

Frama-C: The work in [103] presents that Frama-C could be a platform of static analysis for the programs of C that host the plugin of Impact Analysis. Here, slicing is performed by a plugin, and the analysis of dependency enables developers to visualize the specified variable impact and could run either by GUI or from a line of command. Also, it is meant for utilizing C programs by industry; however, it could be utilized for any size programs for any type of purpose.

The Frama-C also occurs as the only tool in a chosen tools set, which is available open in the form of both open source and commercial tool. Here, community assistance for this equipment could be established well by updating the blog dynamically, tracking the database of bug and wiki.

Impala: The work in [104] presents that the plugin of an eclipse utilizes mining algorithms of data for performing the analysis of the impact on programs of java, before executing modifications. The two program versions are compared; the impala produces graph dependence and forms a set of change for the entire identified modifications and possibly affected entities. The impala tool is optimal to combine with Concurrent Versioning Systems, which enables analysis of the history of the project and understands the complete every class evolution.

The call graphs are generated by a tool for specified programs, and modifications are categorized into distinct kinds like removing/adding methods or classes or modifying classes visibility. Here, the algorithms of the impact are later implemented for attaining impacted changes. The impala endeavours to enhance the confines of former algorithms of static analysis that would not generate optimal outcomes. By concentrating on enhancing algorithms' recall and precision, the maximum accuracy could be attained.

JSlice: The work in [105] presents that JSlice is an eclipse plugin of open source 9, which assists the slicing of dynamic programming. The Virtual Machine of Kaffe utilizes as the backend to retrieve an active slice. Here, criteria specified by a developer are implemented, and outcomes are converted back towards the source code to inspect the UI. Here also the tool provides the flexibility of slicing towards the developer, enabling the option for slicing previous execution or total execution of the specified statement.

The objective of JSlice is to enhance conventional dynamic and static slicing by calculating the related slice. Here, in conventional dynamic slicing, the algorithm gathers the statements that are executed under the criteria of slicing. For computing-related slice, the dependencies were laid in a graph of expanded dynamic dependence. This enables developers to notice statements in the slice, which might impact the criteria of slicing. This could assist greatly in debugging situations where code that is unexecuted might impact the execution of the program.

JUnit/CIA: The work in [26] presents that JUnit/CIA could be a tool of change analysis applied as an increment to JUnit in eclipse. Here, the work in [25] presents that Chianti analysis of the impact tool is utilized for detecting atomic changes of a program, detecting tests impacted by atomic modifications and defining any impacting modifications or every test. Also, this tool could be envisaged to detect the abrupt cause of a test for failing.

The schema for classification is developed to assist change pinpoint, which contributes to test failing. The classifiers are as follows:

- Red: maximum possibility of failure source
- Green: minimum possibility of unsuccessful source
- Yellow: it is between maximum and minimum possibility of inducing unsuccessful source.

The objective of this tool is to detect the modifications which are exactly towards test failing and to label them to be "Red". Here, it grasps the concentration of developers towards the possible problem source.

ImpactViz: The work in [106] presents that ImpactViz could be an eclipse plugin, which enables developers to visualize dependencies of class, involving mined information from SVN. Here, this ability might assist the developer in pinpointing the error of the source and knowing the possible effect of the specified modifications set. Organizing this tool mines the outcomes into impact areas of change and enables the developer to zoom into interesting areas, improved with modifications in history for every definite class.

The tool utilizes call graphs depicting the call graph model and associations among classes for discovery impact preparation. In addition, these graphs form on the basis of coded visualized colour, which forms the significant tool function. The developers might utilize filters for graph trimming towards the required scope and size and communicate with a graph for dependencies of program flow. Here, these features were meant for assisting developers in the procedure of debugging an error. Moreover, integrating this tool with SVN enables the analysis beginning from a recognized bug-free state and traversing through versions until an error source could be discovered.

Fault-Tracer: The work in [107] presents that Fault-Tracer could impact the analysis equipment of change open source 4 applied in the form of plugin aimed at an eclipse. The toll compares the same program of two different versions and recommends regression tests if required. Here, in several ways, the work in [25] presents that Fault-Tracer could be identical to Chianti in objectives and underlying schemes; however, it is claimed for performing better, ranking Chianti's heuristic over 50%.

Fault-Tracer could be made of three views: the view of atomic change, graph view of extended call and view of testing-debugging. The atomic view of change enables developers for viewing and communicating with variances among two chosen versions. Besides, in view of ECG, the developers might see every test of ECG that might assist in understanding the impacting changes. The testing-debugging view exhibits the final outcome of an algorithm of Fault-Tracer, involving the impacted tests and ranks related to atomic modifications for every test. The developer has two distinct project versions in a similar workspace for Fault-Tracer running.

Imp: The work in [108] presents that Imp could be a tool of change analysis, which could be available in the form of a plugin for Versioning System towards program analysing written in C++ or C. The former contribution on the slicing of a static program and cluster dependence is recorded as associated work conducted by plugin Imp. The objective of Imp is to enhance outcomes conducted for addressing the problems of accuracy and performance with the slicing of a static program. The work in [95] presents that CodeSurfer is utilized for implementing the slicing of a static program and is integrated with Version Controller.

The Imp could be utilized for distinct instances like analysis of dependency, analysis of what-if, regression testing and analysis of risk. Highlighting the impacts in an editor of VS and analysis summary could be presented for developers in the pop-up. When it has built-on analysis schemes utilized in CodeSurfer, the Imp claims for delivering enhancements in accuracy and performance.

ImpactMiner: The work in [97] presents that ImpactMiner could be tool 5 of the change impact, which utilizes repository mining of the source code and offers three methods of analysis: retrieval of information, analysis of history and active analysis. As an application of plugin eclipse, ImpactMiner could be examined for java programs, utilizing the SVN integration and observing developer models, which are edited typically in combination with currently detected modifications.

6.4 CHAPTER SUMMARY AND RECOMMENDATIONS

Many of the tools related to Impact Analysis are proposed for applications of java. This was not an unpredicted finding; the java could be a well-known language. As Impact Analysis advantages are realized, it might be an added advantage for the industry of software as a total for assuring that these methods and tools might be used for other languages.

The methods of static analysis are more dominant than the techniques of dynamic analysis. When two types of schemes could be utilized for distinct purposes and towards distinct goals, the dynamic analysis could be recognized typically as capable of providing more precise outcomes in the form of static analysis, because of its contribution to the actual execution of the program. It could be significant to however note that not entire schemes were equally formed and several techniques of static analysis offered accuracy enhancements and performance over former static IA version methods.

The call graph, dependence graphs of a program and slicing of the program are most commonly detected techniques of Impact Analysis in the chosen set of tool. Some changes on the standard scheme offer definite optimizations. Although there were several Impact Analysis methods [85, 91, 109], the absence of a greater variety of methods opens up the risk of all these methods getting lost. The work [85, 86] presents that the tools significantly improve methods for enhancing method utilization.

The outcomes from the review of literature and inspection usability have not covered several encouraging fields of future research. For building on the usability of informal inspection outcomes, this region of work added advantage from a complete analysis of usability, including manifold consumers for fully understanding realistic requirements in a developer.

Some of the additional work might be conducted on detected tools in this book for discovering the true life-span and tool status. It could be possible where the tool might still be proposed dynamically, but not publicized well.

It is also perceived that when several tools are present related to Impact Analysis, there was no Impact Analysis open source for the libraries of API. There were libraries for examining tree 2 syntax abstract of java; however, none of the open source APIs executes the analysis of the impact. Hence, the integration of Impact Analysis methods as open source API improves the scope of impact analysis in software industry.

Probably, it could be apparent that the most significant chance for the enhancement could be the requirement and desire to bridge the gap between the tools of Impact Analysis proposed in academia and the tools that are utilized by an industry of software development. It needs to be an optimal practice for planning and development of the future tool.

7

OBSERVATIONS AND FUTURE RESEARCH POSSIBILITIES

Software applications continue to emerge with increasing complexity and frequent modifications. Accordingly, the importance of evaluating the impact of such modifications on available artefacts is increasing. Hence, Change Impact Analysis has been the focus of the study of most researchers in the software re-engineering field. Consequently, the volume of research works proposed on Change Impact Analysis in contemporary literature is vast. Nevertheless, not many works can serve as initial support for future works on Change Impact Analysis.

The chapter framed a set of research queries and possible options for the future scope of the work, which are mentioned below. Initially, the research study identified the limited availability of empirical evaluation of suggested concepts. In addition, no techniques to serve the entire program development procedure were identified as this needs strong coupling among multiple phases of the process. In addition, most researchers suggested categorization concepts for types of modification and dependency that can affect Change Impact Analysis. As a part of the future scope of Change Impact Analysis research, a systematic investigation needs to be performed considering the pros and cons of the available studies.

7.1 FUTURE STUDY SCOPE

- Limited availability of Change Impact Analysis for heterogeneous artefacts and multi-perspective projects. As discussed in previous chapters, most of the contemporary literature operates in single-perspective projects, and over 65% of the studies operate on the basis of the source code [77]. A few Change Impact Analysis studies are designed to handle multiple artefacts. Nevertheless, these studies have several shortcomings and restrict their usage to single-perspective projects.

- This book reviewed the available literature in the field and observed inconsistencies among different Change Impact Analysis techniques proposed for modelling modification functions. None of the studies could be considered as standard techniques for addressing the Change Impact Analysis operations. As any Change Impact Analysis requires accuracy in the classification of modification functions and then in the next phase requires successful modelling of these operations, a standard evaluation of multiple modification functions is necessary.
- Further, dependency relations resulting in the proliferation of modifications across artefacts also face similar challenges. Dependency approaches suggested in the observed research works are inadequately classified and are not clearly specified. In addition, the classified categories are often not comprehensive and contradict with other categories. Accordingly, much detailed observation of these relations is necessary to ensure acceptable Change Impact Analysis quality.
- The result of this study focusing on Change Impact Analysis, which is crucial objective of this book. Only a small number of techniques are observed to depict programmers of the process and the reason for a given artefact being affected by the suggested modification and the steps to be taken to handle the change. Irrespective of the final use of outcomes of Change Impact Analysis, a clear understanding of the process and the reason for this impact is essential, and all possible impacts should be handled similarly.

REFERENCES

1. Arnold, Robert S., and Shawn A. Bohner. "Impact analysis-towards a framework for comparison." *1993 Conference on Software Maintenance, CSM-93*. IEEE, 1993.
2. Turver, Richard J., and Malcolm Munro. "An early impact analysis technique for software maintenance." *Journal of Software: Evolution and Process* 6.1 (1994): 35–52.
3. Hassine, Jameleddine, et al. "Change impact analysis for requirement evolution using use case maps." *Eighth International Workshop on Principles of Software Evolution*. IEEE, 2005.
4. Shiri, Maryam, Jameleddine Hassine, and Juergen Rilling. "A requirement level modification analysis support framework." *Third International IEEE Workshop on Software Evolvability 2007*. IEEE, 2007.
5. Breech, Ben, et al. "Online impact analysis via dynamic compilation technology." *20th IEEE International Conference on Software Maintenance, 2004. Proceedings*. IEEE, 2004.
6. Law, James, and Gregg Rothermel. "Incremental dynamic impact analysis for evolving software systems." *14th International Symposium on Software Reliability Engineering, 2003. ISSRE 2003*. IEEE, 2003.
7. Bennett, Keith H., and Václav T. Rajlich. "Software maintenance and evolution: A roadmap." *Proceedings of the Conference on the Future of Software Engineering*. ACM, 2000.
8. Bohner, Shawn Anthony. "A graph traceability approach for software change impact analysis.", Dissertation, George Mason University, 1995.
9. Horwitz, Susan, Thomas Reps, and David Binkley. "Inter procedural slicing using dependence graphs." *ACM SIGPLAN Notices* 39.4 (2004): 229–243.
10. Podgurski, Andy, and Lori A. Clarke. "A formal model of program dependences and its implications for software testing, debugging, and maintenance." *IEEE Transactions on Software Engineering* 16.9 (1990): 965–979.

11. Ferrante, Jeanne, Karl J. Ottenstein, and Joe D. Warren. "The program dependence graph and its use in optimization." *ACM Transactions on Programming Languages and Systems* 9 (1987): 319–349.

12. Kama, Nazri, and Faizul Azli Abdul Ridzab. "Requirement level impact analysis with impact prediction filter." *4th International Conference on Software Technology and Engineering (Icste 2012).* 2012.

13. Kama, Nazri, Tim French, and Mark Reynolds. "Considering patterns in class interactions prediction." In *International Conference on Advanced Software Engineering and Its Applications* (2010, December) (pp. 11–22). Springer, Berlin, Heidelberg.

14. Gotel, Orlena Cara Zena. *Contribution Structures for Requirements Traceability.* Dissertation. University of London, London, United Kongdom, 1995.

15. Ali, Hassan Osman, M. Z. Abd Rozan, and Abdullahi Mohamud Sharif. "Identifying challenges of change impact analysis for software projects." *2012 International Conference on Innovation Management and Technology Research (ICIMTR).* IEEE, 2012.

16. Lee, Michelle L. *Change Impact Analysis of Object-Oriented Software.* Fairfax, VA: George Mason University, 1998.

17. Pfleeger, Shari Lawrence, and Joanne M. Atlee. *Software Engineering: Theory and Practice.* Chennai, India: Pearson Education India, 1998.

18. Pfleeger, Shari Lawrence, and Shawn A. Bohner. "A framework for software maintenance metrics." *Conference on Software Maintenance, 1990. Proceedings.* IEEE, 1990.

19. Bohner, Shawn A. "Software change impacts-an evolving perspective." *International Conference on Software Maintenance, 2002. Proceedings.* IEEE, 2002.

20. Lindvall, Mikael, and Kristian Sandahl. "How well do experienced software developers predict software change?." *Journal of Systems and Software* 43.1 (1998): 19–27.

21. Cohen, Jacob. "A coefficient of agreement for nominal scales." *Educational and Psychological Measurement* 20.1 (1960): 37–46.

22. Bohner, Shawn A. "Extending software change impact analysis into COTS components." *27th Annual NASA Goddard/IEEE Software Engineering Workshop, 2002. Proceedings.* IEEE, 2002.

23. Hassaine, Salima, et al. "A seismology-inspired approach to study change propagation." *2011 27th IEEE International Conference on Software Maintenance (ICSM).* IEEE, 2011.

24. Popescu, Daniel. "Impact analysis for event-based components and systems." *2010 ACM/IEEE 32nd International Conference on Software Engineering, Vol. 2.* IEEE, 2010.

25. Ren, Xiaoxia, et al. "Chianti: A tool for change impact analysis of java programs." *ACM Sigplan Notices* 39.10 (2004): 432–448.
26. Stoerzer, Maximilian, et al. "Finding failure-inducing changes in java programs using change classification." *Proceedings of the 14th ACM SIGSOFT International Symposium on Foundations of Software Engineering.* ACM, 2006.
27. Law, James, and Gregg Rothermel. "Whole program path-based dynamic impact analysis." *Proceedings of the 25th International Conference on Software Engineering.* IEEE Computer Society, 2003.
28. Apiwattanapong, Taweesup, Alessandro Orso, and Mary Jean Harrold. "Efficient and precise dynamic impact analysis using execute-after sequences." *Proceedings of the 27th International Conference on Software Engineering.* ACM, 2005.
29. Huang, Lulu, and Yeong-Tae Song. "Precise dynamic impact analysis with dependency analysis for object-oriented programs." *Proceedings of the 5th ACIS International Conference on Software Engineering Research, Management & Applications.* IEEE Computer Society, 2007.
30. Tip, Frank. *A Survey of Program Slicing Techniques.* Amsterdam, Netherlands: Centrum voor Wiskunde en Informatica, 1994.
31. Ranganath, Venkatesh Prasad, and John Hatcliff. "Slicing concurrent Java programs using Indus and Kaveri." *International Journal on Software Tools for Technology Transfer* 9.5–6 (2007): 489–504.
32. The Wisconsin Program-Slicing Tool. http://research.cs.wisc.edu/wpis/html/, May 2014. (Accessed on November, 20th 2014).
33. Girba, Tudor, Stéphane Ducasse, and Michele Lanza. "Yesterday's weather: Guiding early reverse engineering efforts by summarizing the evolution of changes." *20th IEEE International Conference on Software Maintenance, 2004. Proceedings.* IEEE, 2004.
34. Zimmermann, Thomas, et al. "Mining version histories to guide software changes." *IEEE Transactions on Software Engineering* 31.6 (2005): 429–445.
35. Kagdi, Huzefa, et al. "Blending conceptual and evolutionary couplings to support change impact analysis in source code." *2010 17th Working Conference on Reverse Engineering (WCRE).* IEEE, 2010.
36. Canfora, Gerardo, et al. "Using multivariate time series and association rules to detect logical change coupling: An empirical study." *2010 IEEE International Conference on Software Maintenance (ICSM).* IEEE, 2010.
37. Bode, Stephan. Quality Goal Oriented Architectural Design and Traceability for Evolvable Software Systems. Diss. Technische Universität Ilmenau, Germany, 2011.
38. Cavnar, William B., and John M. Trenkle. "N-gram-based text categorization." *Ann Arbor MI* 48113.2 (1994): 161–175.

39. Marcus, Andrian, and Jonathan I. Maletic. "Recovering documentation-to-source-code traceability links using latent semantic indexing." *Proceedings of the 25th International Conference on Software Engineering.* IEEE Computer Society, 2003.
40. Poshyvanyk, Denys, et al. "Using information retrieval based coupling measures for impact analysis." *Empirical Software Engineering* 14.1 (2009): 5–32.
41. Binkley, David, and Dawn Lawrie. "Information retrieval applications in software maintenance and evolution." *Encyclopedia of Software Engineering* P. Laplante, Ed.: Taylor & Francis LLC (2010): 454–463.
42. Sharafat, Ali R., and Ladan Tahvildari. "A probabilistic approach to predict changes in object-oriented software systems." *11th European Conference on Software Maintenance and Reengineering, 2007, CSMR'07.* IEEE, 2007.
43. Ceccarelli, Michele, et al. "An eclectic approach for change impact analysis." (2010).
44. Cabot, Jordi, and Martin Gogolla. "Object constraint language (OCL): a definitive guide." *International School on Formal Methods for the Design of Computer, Communication and Software Systems.* Springer, Berlin, Heidelberg, April 2012.
45. World Wide Web Consortium. *Xml Path Language (xpath) 2.0.* (2010).
46. EMF Query. http://projects.eclipse.org/projects/modeling.emf.query. (Accessed on November, 20th 2014).
47. Briand, Lionel C., Yvan Labiche, and George Soccar. "Automating impact analysis and regression test selection based on UML designs." *International Conference on Software Maintenance, 2002. Proceedings.* IEEE, 2002.
48. Müller, Klaus, and Bernhard Rumpe. "A model-based approach to impact analysis using model differencing." in *Proc. International Workshop on Software Quality and Maintainability (SQM'14), ECEASST Journal,* vol. 65, (2014).
49. Gethers, Malcom, et al. "Integrated impact analysis for managing software changes." *2012 34th International Conference on Software Engineering (ICSE).* IEEE, 2012.
50. Arnold, Robert S. *Software Change Impact Analysis.* Washington, DC, United States: IEEE Computer Society Press, 1996.
51. Anezin, D. Process and Methods for Requirements Tracing (Software Development Life Cycle). Dissertation, George Mason University, (1994).
52. Han, Jun. "Specifying the structural properties of software documents." *Journal of Computing and Information* 1 (1994): 1333–1351.
53. Davis, Jesse, and Mark Goadrich. "The relationship between Precision-Recall and ROC curves." *Proceedings of the 23rd International Conference on Machine Learning.* ACM, 2006.

54. Vallabhaneni, S. Rao. *Auditing the Maintenance of Software*. New Delhi, India: Prentice-Hall, Inc., 1987.
55. Arthur, L. J., *Software Evaluation*. New Jersey, United States: John Wiley and Sons, 1999.
56. Jashki, Mohammad-Amin, Reza Zafarani, and Ebrahim Bagheri. "Towards a more efficient static software change impact analysis method." *Proceedings of the 8th ACM SIGPLAN-SIGSOFT Workshop on Program Analysis for Software Tools and Engineering*. ACM, 2008.
57. Li, Yin, et al. "Requirement-centric traceability for change impact analysis: A case study." *International Conference on Software Process*. Springer, Berlin, Heidelberg, 2008.
58. Amyot, Daniel, and Gunter Mussbacher. "URN: Towards a new standard for the visual description of requirements." *Lecture Notes in Computer Science* 2599 (2003): 21–37.
59. Dahlstedt, Asa G., and Anne Persson. "Requirements interdependencies-moulding the state of research into a research agenda." *The Ninth International Workshop on Requirements Engineering: Foundation for Software Quality (REFSQ 2003)*, Klagenfurt/Velden, Austria, 2003.
60. Lucia, De. "Information retrieval models for recovering traceability links between code and documentation." *International Conference on Software Maintenance, 2000. Proceedings*. IEEE, 2000.
61. Rohatgi, Abhishek, Abdelwahab Hamou-Lhadj, and Juergen Rilling. "An approach for mapping features to code based on static and dynamic analysis." *The 16th IEEE International Conference on Program Comprehension, 2008. ICPC 2008*. IEEE, 2008.
62. Spanoudakis, George. "Plausible and adaptive requirement traceability structures." *Proceedings of the 14th International Conference on Software Engineering and Knowledge Engineering*. ACM, 2002.
63. O'Neal, James S. "Analyzing the impact of changing requirements." *Proceedings of the IEEE International Conference on Software Maintenance (ICSM'01)*. IEEE Computer Society, 2001.
64. Adams, Bram, et al. "The evolution of the linux build system." *Electronic Communications of the EASST* 8 (2008).
65. Seo, Hyunmin, et al. "Programmers' build errors: A case study (at google)." *Proceedings of the 36th International Conference on Software Engineering*. ACM, 2014.
66. Kerzazi, Noureddine, Foutse Khomh, and Bram Adams. "Why do automated builds break? an empirical study." *2014 IEEE International Conference on Software Maintenance and Evolution (ICSME)*. IEEE, 2014.
67. McIntosh, Shane, et al. "Mining co-change information to understand when build changes are necessary." *2014 IEEE International Conference on Software Maintenance and Evolution (ICSME)*. IEEE, 2014.

68. Xia, Xin, et al. "Cross-project build co-change prediction." *2015 IEEE 22nd International Conference on Software Analysis, Evolution and Reengineering (SANER)*. IEEE, 2015.
69. Fluri, Beat, and Harald C. Gall. "Classifying change types for qualifying change couplings." *14th IEEE International Conference on Program Comprehension, 2006. ICPC 2006*. IEEE, 2006.
70. Gall, Harald C., Beat Fluri, and Martin Pinzger. "Change analysis with evolizer and changedistiller." *IEEE Software* 26.1 (2009): 26.
71. Fluri, Beat, et al. "Change distilling: Tree differencing for fine-grained source code change extraction." *IEEE Transactions on Software Engineering* 33.11 (2007): 725–743.
72. Hattori, Lile P., and Michele Lanza. "On the nature of commits." *Proceedings of the 23rd IEEE/ACM International Conference on Automated Software Engineering*. IEEE Press, 2008.
73. Ibrahim, Walid M., et al. "Should I contribute to this discussion?" *2010 7th IEEE Working Conference on Mining Software Repositories (MSR)*. IEEE, 2010.
74. Knab, Patrick, Martin Pinzger, and Abraham Bernstein. "Predicting defect densities in source code files with decision tree learners." *Proceedings of the 2006 International Workshop on Mining Software Repositories*. 2006.
75. Romano, Daniele, and Martin Pinzger. "Using source code metrics to predict change-prone java interfaces." *2011 27th IEEE International Conference on Software Maintenance (ICSM)*. IEEE, 2011.
76. Giger, Emanuel, Martin Pinzger, and Harald C. Gall. "Comparing fine-grained source code changes and code churn for bug prediction." *Proceedings of the 8th Working Conference on Mining Software Repositories*. ACM, 2011.
77. Lehnert, Steffen. "A review of software change impact analysis." Ilmenau University of Technology, Ilmenau, Germany, *Technical Report* (2011).
78. Hammad, Maen, Michael L. Collard, and Jonathan I. Maletic. "Automatically identifying changes that impact code-to-design traceability." *IEEE 17th International Conference on Program Comprehension, 2009, ICPC'09*. IEEE, 2009.
79. Kotonya, Gerald, and John Hutchinson, G. Kotonya and J. Hutchinson, "Analysing the impact of change in COTS-based systems." *Lecture Notes in Computer Science* 3412 2005: 212–222.
80. Lock, Safoora and Shakil Khan Simon. "Concern tracing and change impact analysis: An exploratory study." *ICSE Workshop on Aspect-Oriented Requirements Engineering and Architecture Design, 2009. EA'09*. IEEE, 2009.
81. Ibrahim, Suhaimi, et al. "A requirements traceability to support change impact analysis." *Asian Journal of Information Technology* 4.4 (2005): 345–355.

82. Ibrahim, S., N. B. Idris, M. Munro, and A. Deraman. A software traceability validation for change impact analysis of object oriented software. *Proceedings of the International Conference on Software Engineering Research and Practice & Conference on Programming Languages and Compilers, SERP 2006*, volume 1, pages 453–459, Las Vegas, Nevada, USA, June 2006.

83. Goeritzer, Robert. "Using impact analysis in industry." *2011 33rd International Conference on Software Engineering (ICSE)*. IEEE, 2011.

84. de Souza, Cleidson, and David Redmiles. "An empirical study of software developers' management of dependencies and changes." *2008 ACM/IEEE 30th International Conference on Software Engineering*. IEEE, 2008.

85. Li, Bixin, et al. "A survey of code-based change impact analysis techniques." *Software Testing, Verification and Reliability* 23.8 (2013): 613–646.

86. Tóth, Gabriella, et al. "Comparison of different impact analysis methods and programmer's opinion: An empirical study." *Proceedings of the 8th International Conference on the Principles and Practice of Programming in Java*. ACM, 2010.

87. Weiser, M., Program slicing, *ICSE '81: Proceedings of the 5th International Conference on Software Engineering*, 439–449, IEEE Press, Piscataway, NJ, USA, 1981.

88. Silva, Josep. "A vocabulary of program slicing-based techniques." *ACM Computing Surveys (CSUR)* 44.3 (2012): 12.

89. Harman, Mark, Margaret Okunlawon, Bala Sivagurunathan and Sebastian Danicic. "Slice-based measurement of coupling." *19th ICSE, Workshop on Process Modeling and Empirical Studies of Software Evolution*, Boston, Massachusetts, USA, May 1997.

90. Briand, Lionel C., Jurgen Wust, and Hakim Lounis. "Using coupling measurement for impact analysis in object-oriented systems." *Proceedings IEEE International Conference on Software Maintenance-1999 (ICSM'99).'Software Maintenance for Business Change' (Cat. No. 99CB36360)*. IEEE, 1999.

91. Harman, Mark, and Robert Hierons. "An overview of program slicing." *Software Focus* 2.3 (2001): 85–92.

92. Zanjani, Motahareh Bahrami, George Swartzendruber, and Huzefa Kagdi. "Impact analysis of change requests on source code based on interaction and commit histories." *Proceedings of the 11th Working Conference on Mining Software Repositories*. ACM, 2014.

93. Lyle, James R., et al. "Unravel: A CASE tool to assist evaluation of high integrity software. Volume 1: Requirements and design." National Institute of Standards and Technology, Computer Systems Laboratory, Gaithersburg, MD, 1995.

94. Venkatesh, G. A. "Experimental results from dynamic slicing of C programs." *ACM Transactions on Programming Languages and Systems (TOPLAS)* 17.2 (1995): 197–216.

95. Teitelbaum, Tim. "Codesurfer." *ACM SIGSOFT Software Engineering Notes* 25.1 (2000): 99.

96. Bilal, Haider, and Sue Black. "Using the ripple effect to measure software quality." *International Conference on Software Quality Management*. Vol. 13, Cheltenham, Gloucestershire, UK, 2005.

97. Dit, Bogdan, et al. "Impactminer: A tool for change impact analysis." *Companion Proceedings of the 36th International Conference on Software Engineering*. ACM, 2014.

98. Lattix Inc. *Lattix*, 2017. http://lattix.com/. (Accessed on July, 11th 2018)

99. Ranganath, Venkatesh Prasad, and John Hatcliff. "An overview of the indus framework for analysis and slicing of concurrent java software (keynote talk-extended abstract)." *2006 Sixth IEEE International Workshop on Source Code Analysis and Manipulation*. IEEE, 2006.

100. Jayaraman, Ganeshan, Venkatesh Prasad Ranganath, and John Hatcliff. "Kaveri: Delivering the indus java program slicer to eclipse." *International Conference on Fundamental Approaches to Software Engineering*. Springer, Berlin, Heidelberg, 2005.

101. Buckner, Jonathan, et al. "JRipples: A tool for program comprehension during incremental change." *13th International Workshop on Program Comprehension (IWPC'05)*. IEEE, 2005.

102. Coder Gears. *Jarchitect*, 2017. http://www.jarchitect.com. (Accessed on August, 27th 2018).

103. Cuoq, Pascal, et al. "Frama-c." *International Conference on Software Engineering and Formal Methods*. Springer, Berlin, Heidelberg, 2012.

104. Hattori, Lile, et al. "On the precision and accuracy of impact analysis techniques." *Seventh IEEE/ACIS International Conference on Computer and Information Science (ICIS 2008)*. IEEE, 2008.

105. Wang, Tao, and Abhik Roychoudhury. "Using compressed bytecode traces for slicing Java programs." *Proceedings of the 26th International Conference on Software Engineering*. IEEE Computer Society, 2004.

106. Follett, Matthew, and Orland Hoeber. "ImpactViz: Visualizing class dependencies and the impact of changes in software revisions." *Proceedings of the 5th International Symposium on Software Visualization*. ACM, 2010.

107. Zhang, Lingming, Miryung Kim, and Sarfraz Khurshid. "FaultTracer: A change impact and regression fault analysis tool for evolving Java programs." *Proceedings of the ACM SIGSOFT 20th International Symposium on the Foundations of Software Engineering*. ACM, 2012.

108. MAcharya, Mithun, and Brian Robinson. "Practical change impact analysis based on static program slicing for industrial software systems." *Proceedings of the 33rd International Conference on Software Engineering.* ACM, 2011.
109. Lehnert, Steffen. "A taxonomy for software change impact analysis." *Proceedings of the 12th International Workshop on Principles of Software Evolution and the 7th Annual ERCIM Workshop on Software Evolution.* ACM, 2011.

INDEX